IF YOU LIKE
Butterflies

For Tavi —LMS

For Żak and Jess —GS

The author would like to thank Chris Hartley and Michael Toliver for their thoughtful evaluation and input on this project.

Chris Hartley, Science Education Coordinator, The Sophia M. Sachs Butterfly House, Chesterfield, MO 63017, www.butterflyhouse.org

Michael Toliver, Professor Emeritus, Biology (PhD in Entomology), Eureka College, Eureka, IL 61530 (Former Secretary of the Lepidopterists' Society)

About This Book

The illustrations for this book were done in water-based inks on hand-pressed paper. This book was edited by Christy Ottaviano and designed by Tracy Shaw. The production was supervised by Kimberly Stella, and the production editor was Jen Graham. The text was set in Cronos Pro, and the display type was hand-lettered by Åsa Gilland.

IF YOU LIKE
Butterflies

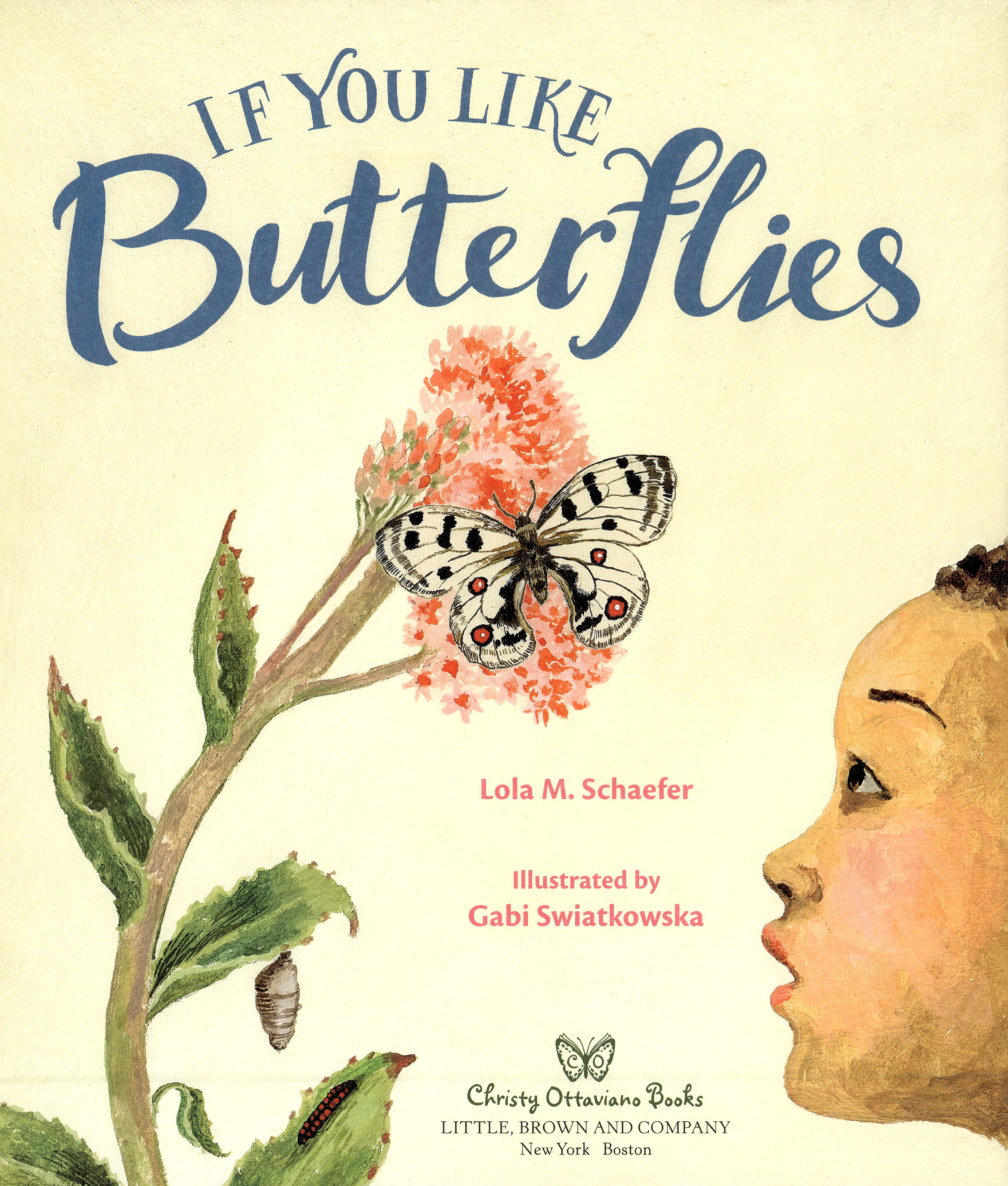

Lola M. Schaefer

Illustrated by
Gabi Swiatkowska

Christy Ottaviano Books
LITTLE, BROWN AND COMPANY
New York   Boston

If **you like butterflies,**
you probably read about them,
visit butterfly houses,
or watch them flit and flutter,
rise and glide,
near your home.

Every kind of butterfly is different.
One species is large—super large—
another is tiny,
and all the rest fall somewhere in between.

Some have bright colors on their wings,
others soft.
And every species has its own unique markings.

Queen Alexandra's Birdwing

Western Pygmy Blue

Guava Skipper

Two-eyed Eighty-eight

Priam's Birdwing

Chequered Skipper

Great Blue Hookwing

California Dogface

Yellow Pansy

Small Copper

Common Jester

Old World Swallowtail

Fluminense Swallowtail

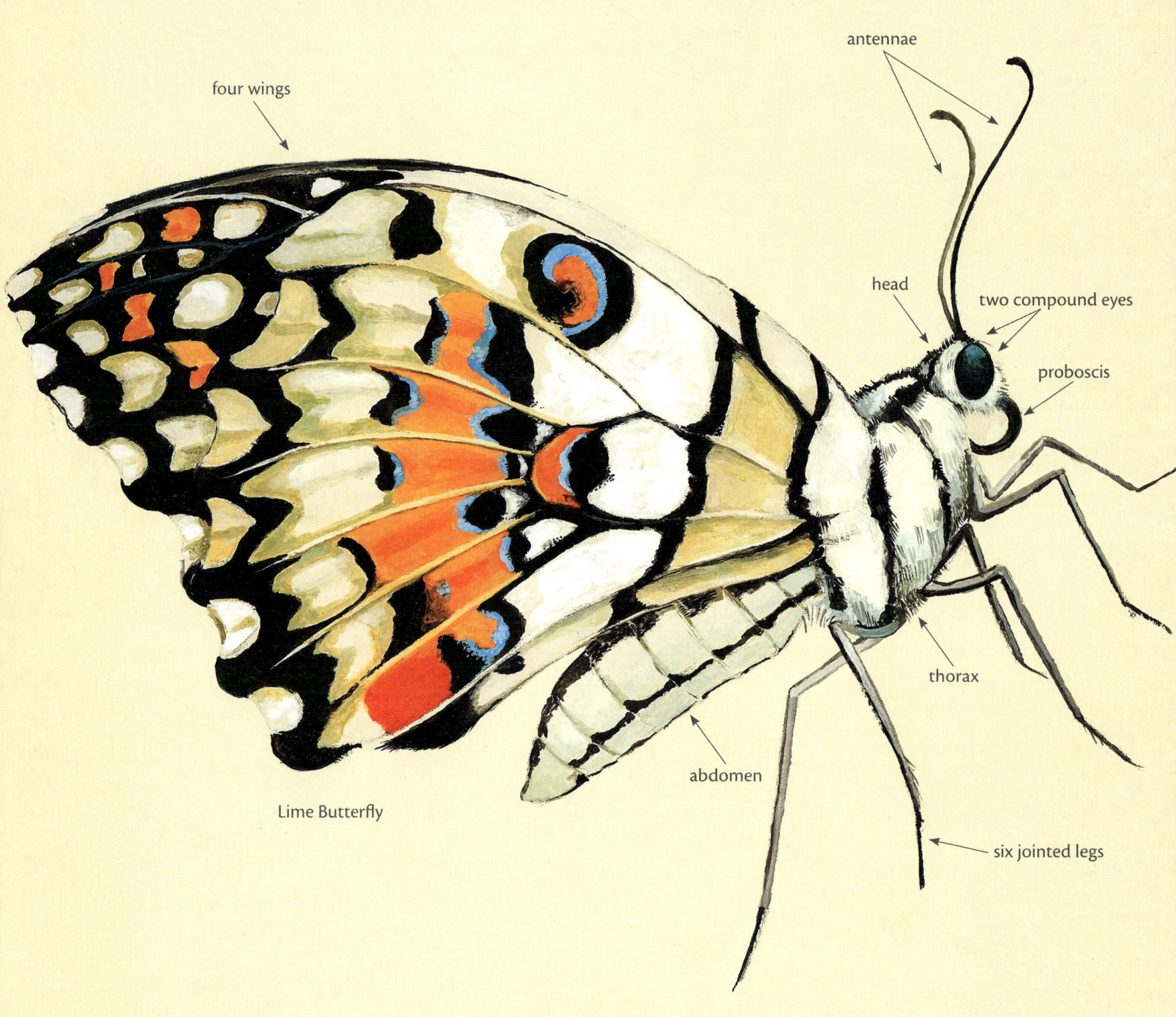

four wings

antennae

head

two compound eyes

proboscis

thorax

abdomen

six jointed legs

Lime Butterfly

If you study butterflies,
you know that they are insects with
six jointed legs,
a head,
a thorax,
an abdomen,
antennae,
compound eyes,
a proboscis,
and wings . . .

wings that carry them from flower to flower,
morning through late afternoon,
sipping nectar
and spreading pollen.

When a female butterfly is not eating,
she is searching for nutritious host plants
for her eggs.
When she tastes the correct plant,
she lays an egg,
or a group of eggs,
on the underside of a leaf
or on the top.
Sometimes a female butterfly hides her eggs
on dry grass stems,
dead leaves,
even on soil.

Pipevine Swallowtail

Butterfly eggs can be round or oval,
smooth or ribbed,
shaped like a dome
or a barrel.
Most are yellow,
white,
green,
or brown,
but some are almost transparent.

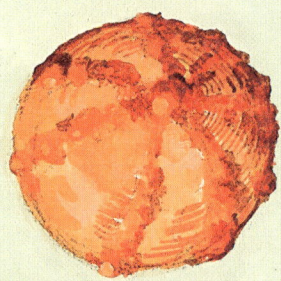

Pipevine Swallowtail

Atala

Zebra Swallowtail

Black Swallowtail

Cabbage White

Banana Skipper

Master's Skipper

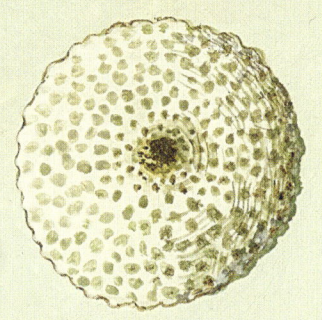

Lupine Blue

American Lady

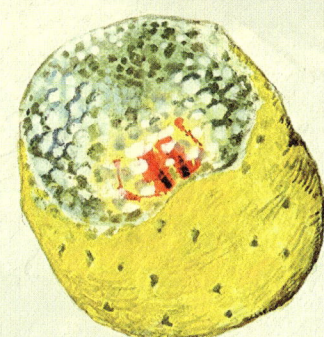

Speckled Wood

Morpho cisseis phanodemus

Clouded Yellow

Yellow Vein Lancer

Mourning Cloak

Blue Morpho

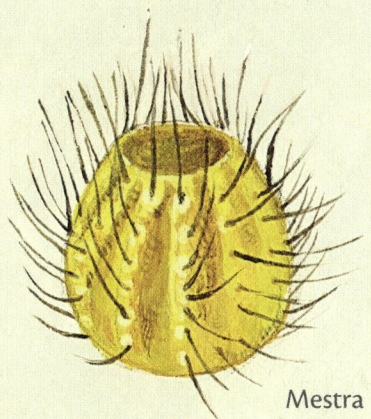

Mestra

Common Blue

Wood White

Brimstone

If you find a butterfly egg,
return daily until it begins to change color.
The larva, or caterpillar, is forming.
If you're patient,
you will see the caterpillar chew a tiny hole
through the egg and crawl out.

Plain Tiger

These caterpillars look nothing like butterflies.
And since they are tender
and bite-size,
they need to defend themselves from predators with . . .

long hair or fur that can cause a rash,
spikes or horns,
camouflage,
bad taste,
markings that confuse,
or a foul smell.

Blue Morpho

Spicebush Swallowtail

Common Nawab

Pipevine Swallowtail

Giant Swallowtail

Monarch

False Apollo

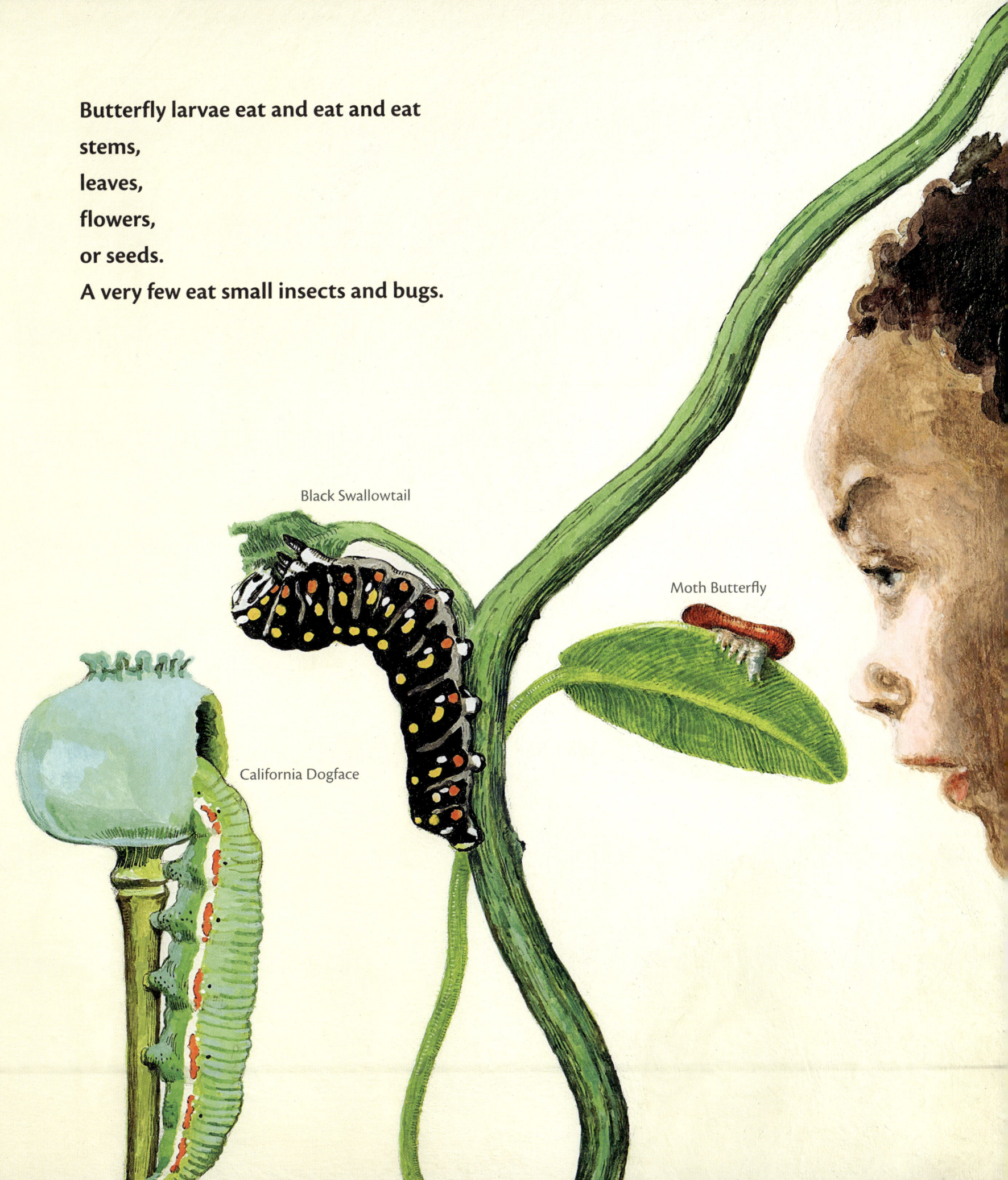

Butterfly larvae eat and eat and eat
stems,
leaves,
flowers,
or seeds.
A very few eat small insects and bugs.

Black Swallowtail

Moth Butterfly

California Dogface

As a larva eats, it grows.
As it grows, it molts,
shedding old skin
for newer, larger skin
again and again
until the final molt,
when the skin hardens and becomes
a chrysalis.

If you discover a chrysalis,
or a pupa,
it could be smooth or spiny,
knobby or wrinkly.
It could even look like a dried leaf.

White Admiral

Ismenius Tiger

Blue Morpho

Ulysses

Cloudless Sulphur

Tawny Coster

Black-veined White

No matter what it looks like on the outside,
inside, the caterpillar is slowly transforming.
Wings emerge.
Antennae appear.

The mouth of the caterpillar becomes the strawlike proboscis.
And as the butterfly forms,
the chrysalis takes on new colors.

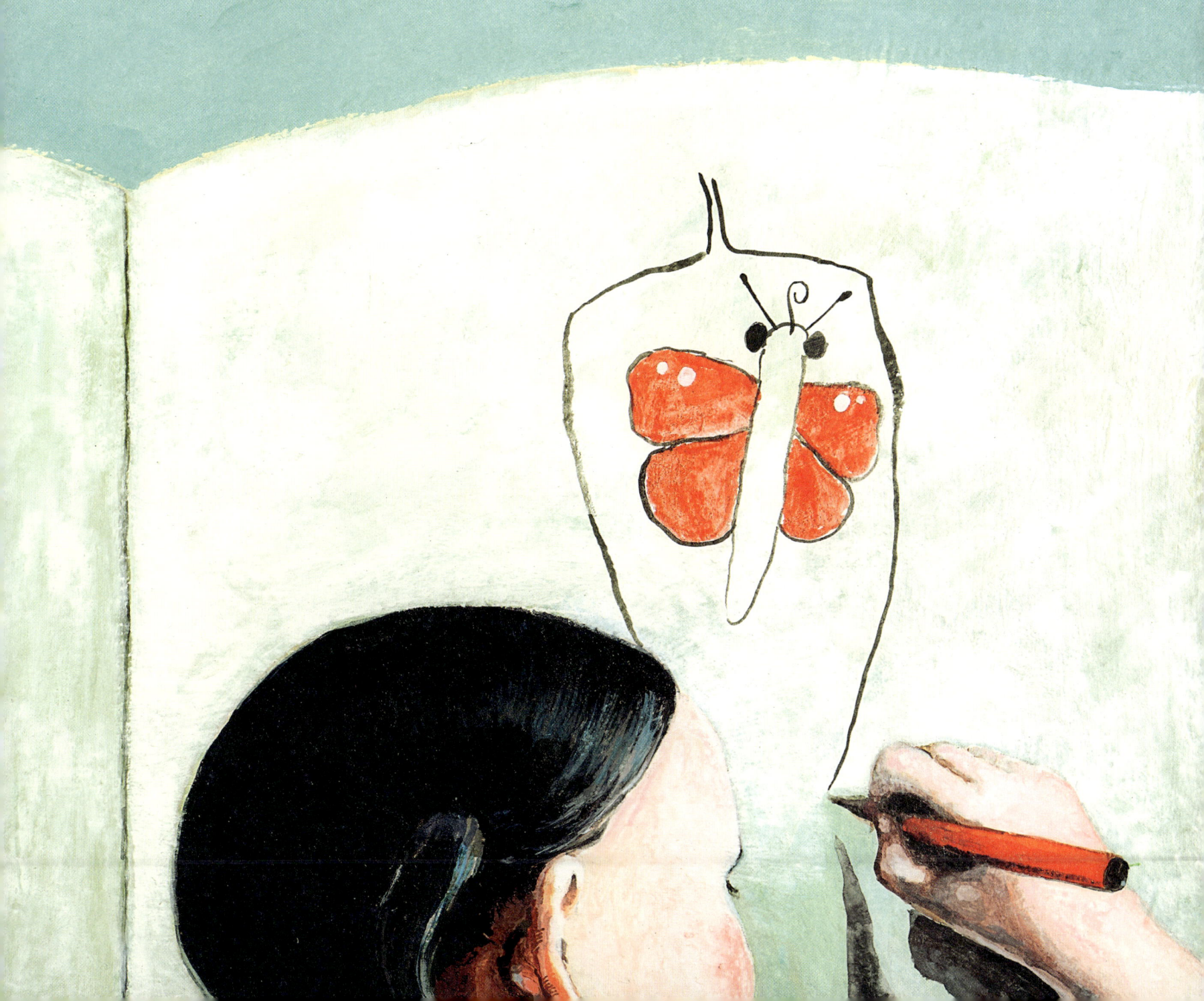

When the butterfly is fully grown,
it splits the pupal shell open.
Using its feet, the butterfly pulls itself out.

Plain Tiger

At first its wings are small and wet.
Soon they fill with body liquids,
making them longer, stronger.
The butterfly flaps its wings as they dry until . . .

it lifts into the air
and flies to its first meal of nectar.

If you watch butterflies,

you know that they are out and about in the warmer months.

When the temperatures get colder,

some butterflies hibernate.

They find protection behind the bark of trees,

in thick plants,

beneath the wood in old fences,

or sometimes in basements or attics.

More often, it's the eggs, caterpillars, or chrysalises that remain

dormant through the winter.

Many butterflies migrate
in fall and spring,
flying from one warm place to another.

Monarch

They travel hundreds,
sometimes thousands,
of miles,
and along the way they sip nectar.

If you like butterflies,
you might grow some of their favorite plants
in your yard,
around your neighborhood,
or along the roads.
And you will see butterflies
flit and flutter,
rise and glide,
year after year after year.

A Butterfly's Life Cycle

A butterfly usually lays eggs on stems or leaves near the food that the larva will need.

The caterpillar, or larva, emerges from the egg and eats the food, growing larger and shedding its skin several times.

The chrysalis, or pupa, is formed by the caterpillar during its final molt. Inside, parts of the caterpillar dissolve into a liquid, leaving the front six legs. From some of this liquid, eyes, wings, and antennae form, and a butterfly takes shape.

The adult butterfly, or imago, pushes out of the chrysalis. It will eat nectar and look for a plant on which to lay its eggs.

Flying Between Two Homes: Butterfly Migration

A few species of butterflies, such as the Monarch, Painted Lady, and Cloudless Sulphur, fly great distances twice each year. This is called migration. In autumn, migrating butterflies leave northern homes and fly south. These different species migrate because their northern homes get too chilly.

Butterflies are cold-blooded insects. Most cannot survive in temperatures that drop to freezing or below. Also, their food supply—the nectar of flowers—disappears in the winter when the plants die back. Butterflies need nectar to live.

Every year, near October, butterflies who emerged in late summer or early fall migrate south. They gather in large groups, sometimes in the thousands, and fly 1,500 to 3,500 miles. Some species fly all the way to Southern California or Mexico. Others do not travel quite that far.

Depending on the species and the direction of the warm air currents, butterflies fly anywhere from 100 to 11,000 feet above the ground. At the higher altitudes, the wind speeds increase. If the butterflies have a good tailwind, they can fly much farther.

The butterflies that migrated remain in their wintering homes until it is time to lay eggs in the spring. Then they fly back north to deposit eggs on their larval (host) plants.

Since most of these butterflies have a lifespan of one to eight months, it is never the same butterfly that returns the next winter to the warmer locations. Instead, it is usually the third or fourth generation of those butterflies that makes the trip. And yet . . . Monarch butterflies return to the same trees in the same location year after year.

Planting a Butterfly Garden

You can plant a butterfly garden that will not only provide food and shelter to these lovely insects but also bring you hours of relaxing entertainment. You can find many resources that will help you build a garden suited to the butterflies in your specific location. First you need host plants, those on which butterflies lay their eggs so the newly hatched caterpillars have the food they need to grow. Depending on where you live and the kinds of butterflies that live in your region, the host plants can vary. Some of the most common ones are:

fennel	dill	alfalfa	sunflower
milkweed	hollyhock	mallow/hibiscus	pansy
aspen tree	apple tree	locust tree	clover

You might want to plant these hosts toward the back of your garden or yard since the plants will be damaged once the caterpillars emerge and eat, eat, eat. Next you will need to grow nectar plants, those flowers that provide food for the butterflies. Again, find out which of these is native to your area and will grow most easily:

marigold	butterfly bush	butterfly weed	lantana
blue cardinal flower	zinnia	joe-pye weed	coneflower
dandelion	oregano	mistflower	cosmos

It's best to plant your garden in a sunny spot—a place that receives at least six hours of sunlight each day. It's also helpful if your garden is sheltered from strong winds. Butterflies get their required water from nectar. However, they need salts and minerals to remain healthy. Sink shallow pans or dishes into the ground between the different plants. Fill them partway with dirt and sand, and place a few rocks on top so butterflies can land. Keep the soil and sand moist and have an adult add a small amount (½ teaspoon) of ammonia to your "puddle." The butterflies will come!

Read More

Burris, Judy, and Wayne Richards. *Butterflies of North America*. Minocqua, WI: Willow Creek Press, 2016.

Davidson, Lauren. *Butterflies for Kids: A Junior Scientist's Guide to the Butterfly Life Cycle and Beautiful Species to Discover*. New York: Rockridge Press, 2021.

Heiligman, Deborah. *From Caterpillar to Butterfly*. Let's Read and Find Out. New York: HarperCollins Publishers, 2015.

Latimer, Jonathan P., and Karen Stray Nolting. *Butterflies*. Peterson Field Guides for Young Naturalists. Boston: Houghton Mifflin Harcourt, 2000.

Pryor, Katherine. *Home Is Calling: The Journey of the Monarch Butterfly*. New York: WorthyKids, 2023.

Websites to Visit

Butterflies and Moths of North America	**www.butterfliesandmoths.org**
The Children's Butterfly Site	**www.kidsbutterfly.org**
Monarch Watch	**www.monarchwatch.org**
North American Butterfly Association	**www.naba.org**